Who Turns The Light?

by

Lance M. Osadchey

First published by AuthorHouse 04/07/04

ISBN: 1-4140-8071-9 (e-book)
ISBN: 1-4184-1383-6 (Paperback)

Library of Congress Control Number: 2004091610

Printed in the United States of America
Bloomington, IN

This book is printed on acid free paper.

Dedicated to:

My brother Bruce and his wife Carol

My son Mark and his wife Karen

My daughter Tanya and her husband Sean

My daughter Kerstin

My grandchildren Miranda, Casey Jack, Trevor, Savanna, Reilly, and Phoebe

My puppy Shadow

Steve

A local astronomer

Physlink Friends

Friends at !st Books

Special thanks to Tanya for proofreading the first galley

And to all the skeptics on the road to this story:

Authors, scientists, university personnel, scientific publishers, space researchers, etc., etc., who often never considered my point of view nor answered some basic questions, never answered e-mails or letters. This excludes the local astronomer who was very helpful, though in the camp of "it is impossible".

Table of Contents

Setting

This is a story of a scientific endeavor, which is being told to entertain people interested in how the world works. Also, it is intended for people that know the subject in very good detail. It is my story and to paraphrase a prominent voice you can believe it or not believe it. It is my story and I will tell it as an interview. Think of the literary style as a two-person play where I play both people. The questions are questions I would want asked to explain my point of view and work. Some of the questions are based on actual questions posed to me by the relative few I have talked with about this topic. It is perhaps the start of a better understanding of the world in which we live. It is not meant to downplay prominent scientists in the past or present. These men and women are my heroes and if anything, this idea is to enhance the work they have done. It is not a challenge to any present theory but instead bits of science that may aid to develop further ideas.

Reporter:
I am at the home of the author, a large house with minimal furniture. Spartan would be more accurate. His dog,

Shadow, lies on a rug appearing to try to sleep, but he is listening intently. There is a beautiful view of distant mountains and a very large lawn visible from the windows. Now the distant mountains are a green color with a bright blue-sky overhead. Rugged mountain peaks show up in the far distance with bumpy rolling ridges in between those distant mountains and the house.

Introduction

Reporter:

(Turning to the author)

So what prompted you to investigate this topic?

Author:

I am the type of person who enjoys puzzles, such as Rubik's cube or how to play chess or coin weighing problems. I think Nature is a puzzle maker and it is up to people to solve the puzzles. Like a magic trick---how is it done? I use the solving aspect as entertainment and get enjoyment out of solving a puzzle. Not crosswords or regular picture puzzles however. Solving a difficult problem is a very enjoyable pastime for me and now that I have retired I have more time for things such as this. And, I have to say, this particular topic has been a very difficult problem. From my first thoughts while reading Lloyd Swenson's book about Michaleson and Morley, right up to the present.

Using the principles and setup of the experiment that I have completed should provide many different experiments to continue the clarification of exactly what Nature is trying to tell us about the real world. This principle of evaluating motion should lead to many new and unexpected facts and understandings.

Background

Reporter:

How did you become interested in this topic "Who turns the Light?"

Author:

I was waiting at a university library to pick my daughter up for summer break when I saw a book and checked it out. It was "The Measurement of the Velocity of Light" by Michaleson and Morley, written back in, or about, 1860. It was a beautiful scientific endeavor. This was an experiment with great ideas and elaborate equipment, measuring the velocity of light. Quite an accomplishment, it is wonderful science.

To follow up on this topic I found and read the book "The Ethereal Aether" by Lloyd Swenson. In this book, the author chronicled the experiment Michaleson and Morley did to define the Aether, as it was called then. This was a substance proposed by the scientists of that day to explain how light moved through space. In their experiment on that topic, Michaleson and Morley wanted to quantify and describe the Aether and to see if they could prove it existed. The results were null, which means there was not enough evidence to prove the existence of the Aether.

From reading this book, I learned that their experiment to measure for the velocity of light was based on earlier work by Fizzeau. Michaleson and Morley refined that previous experimental set up. It was a bit disillusioning for me to discover that Michaleson and Morley didn't think of the entire set-up themselves. I thought they had. Anyway, in the book I found some questions that did not make sense to me.

Reporter:
What questions?

Author:
To popularize their experiment and to raise money, they used an analogy to make their proposal easier to understand. They used the swimmer and stream analogy about a swimmer who swims two equal distances in a stream. One course goes across the river and back and the other course goes down the river and back to the starting

point. If the river is not flowing, the times of the swim will be identical both ways. If there is a current, the time across and back will be less than the trip down and back, despite the distances of the trip being the same.

I have included the calculations for you so that you can incorporate this information in the interview. Perhaps put them at the end of our talk. That would aid anyone interested in seeing how the times are computed.

Similarly, they formed this analogy for their experiment using two beams of light. Both beams traveled the same distance, but one beam traveled at right angles to the other. The beams were sent to mirrors and then reflected back to a point where the beams were compared.

Reporter:
Compared?

Author:
Yes. The process of interference of two light beams can be used to compare individual light beams, to see if they have the same phase or not. A beam of light can be considered

to travel in a wave-like formation. If two beams of light are sent an equal distance, the waves of each beam can combine to form one wave. If the beams are in phase the light will be bright, the result of the two combining 'in phase'. If the beams combine in exactly 'out of phase' a dark area will appear, representing the light beams canceling each other. The first is called being constructive interference and the latter is called destructive interference. If they do have the same phase, then the beams are considered to have traveled the same distance. Inspecting the phase patterns and determining if there is the same number determines whether the distance is the same or not. If one beam gets out of phase, it is considered to have a different 'travel distance' than the other beam. Since light travels at the same velocity, then the time for travel could be substituted for the distance traveled.

Reporter:
So get back to your question this experiment raised for you.

Author:

Michaleson and Morley reasoned if there was an Aether, it would act as the river in the analogy. If the Aether existed and two light beams at right angles to one another were to pass through it, then as in the river analogy, one beam should move differently than the other as the apparatus was turned in a circle. Light rays passing through the Aether at right angles would have different speeds of transmission and they would find the beams coming back out of phase as they rotated the apparatus. Since it was known the earth is moving in space, they should nevertheless find a phase shift in the two light beams. They found a small shift but not enough to prove the existence of the Aether and thus the results of the experiment were considered null. This was considered the death of the Aether concept.

Reporter:
And your question was?

Author:
Consider the swimmer going across the river in Michaleson and Morley's analogy. If the river is not moving, then the swimmer can go straight across and straight back. However, if the river is flowing, then to hit the point straight across, the

swimmer has to turn and swim slightly into the current to hit the point. On the way back, the swimmer has to again angle into the current to arrive back at the starting point.

Reporter:
So?

Author:
As I was standing on the bank of the Connecticut River, I imagined I was shining a beam of light to the other side. If mirror on the other side of the river was exactly at a right angle to me, it would be struck by the beam and return the beam to me. Now here I have to digress and talk about how light is known to move.

Reporter:
OK go on.

Author:
Light is energy and unlike matter. From what I know of light, it travels straight ahead at all times, unless it is interfered with. It has no knowledge of where it is going, or what lies ahead or behind it. Once initiated, light moves straight

ahead. I have read that light exists in its own 'space-time'. Matter, on the other hand, is different. The particles matter is composed of will pick up and carry motion in the dimensions the particles exist in. Consider a beam of light and an arrow shot at exactly a right angle from a car moving on a straight road. The arrow will pick up the motion of the car and move in the direction the car is going, as well as moving straight at the right angle it was shot at.

A beam of light will not pick up the car's motion. Once released the light will move straight at a right angle to the car at the speed of light. Of course, I am referring to a single ray of light such as a laser beam. Light from a source such as the sun or a star or a light bulb moves in an ever-expanding sphere.

Reporter:

Wait. So you say that light released from a car, at right angles to the movement of the car, will proceed ahead and hit any object that lies on its path. However, an arrow will not hit straight ahead but move down the road with the car's velocity, as well as its own velocity, at right angles to the car's path.

Author:

Yes, that is the way I understand it.

Reporter:

Go back to your question.

Author:

Well, we know the earth is in motion. It moves to the east in rotation giving us day and night. It moves around the sun providing us seasons and the count of a year. The tilt of the earth and this motion combine for the change of seasons. Meanwhile, the sun is moving in our Milky Way galaxy completing a round trip in about 240,000 years. The sun drags the earth and the other planets in our solar system with it.

Getting back to the river analogy: Now consider that we ARE NOT moving. I could send a beam of light straight across the river at 90 degrees to the river edge and it could hit a mirror on the opposite side and bounce back to me.

Now consider we ARE moving. I could not imagine how I could send a beam of light straight across the river and have it hit the mirror. That is, straight across at exactly a 90-degree angle. During the time it took for the light to go across, the mirror would have moved slightly out of the way and the beam would miss the mirror.

These ideas, concepts, are illustrated in the website

laqu@bravepages.com

Reporter:
Hold on a minute. You say if both you and the mirror are in motion, the light beam would not be able to hit the mirror if directed at 90 degrees straight towards the mirror?

Author:
Yes. Consider the situation where there is no motion of the mirror or sender. The light goes over, hits the mirror, and returns to the sender. Now think that the mirror and sender are moving. The light goes over and misses the mirror since it moved from its original position. Even if the light beam struck the mirror and the light was reflected, the sender now

has moved and would not be hit. Of course if the angles the sender used, and if the mirror was aimed at a bit more than 90 degrees, the light could return.

I am assuming the mirror is a small mirror, say just big enough to send the light back with no motion involved. If I were to hit the moving mirror, I would need to direct the light beam slightly ahead of its actual position. Remember that its actual position is not where one sees it because as the light comes from the mirror everything has moved slightly during the transit time of the light. Of course I am talking about minor movement, as the distances are rather small and light moves rapidly.

Reporter:

OK. How about this? Suppose you are not moving on one side of the river and the mirror is moving along the other bank. The mirror starts on your left and moves continuously toward the right. It is pointed at 90 degrees toward you. As it moves, it passes a point directly across from you, then continues on. Are you saying that if your light beam is shining straight across the river, the mirror will not be struck and return the light?

Author:

If I am not moving and I send a light beam straight across at 90 degrees, at the moment the mirror is exactly opposite me, then the beam will hit the mirror and bounce straight back to me. At the moment the light leaves me, at that moment it starts its journey, the mirror is a little to the left of the exact opposite side. When the ray moves across the river the mirror continues to move to the right and when it is exactly 90 degrees across, it hits the mirror and the mirror sends it straight back. Since I am not moving, it does hit me. This is similar to the situation where both the mirror and I are not moving, but only at that brief instance.

Reporter:

I have to think about that. I need a break. Do you have anything to drink?

A bit later.

Reporter:

Thanks. Now what was your point?

Author:

Back to the river analogy:

Let's go over it again. Consider the case where there is no motion involving the mirror or me. Both the mirror and myself are at exactly 90 degrees opposite each other. That is, exactly straight across from one another. If I were going to hit the mirror and have it return the light to me, I could send it straight at 90 degrees across the river and back, much like the swimmer who swam straight across and back when there was no current in the river. Again, if there were no motion, yes I could hit the opposite mirror at a 90-degree aim.

However, if the mirror and I were in motion, I could not send it at 90 degrees and hit the mirror. The mirror would have time to move out of the way as I sent the light across and the light would then miss the mirror.

Since I know the swimmer had to angle his body slightly up river to hit his point straight across when there was a current. And since Michaleson and Morley did get a return of their light on a mirror DESPITE both the source and mirror moving, I had to ask "Who turned the Light?" to allow it to hit the mirror.

Reporter:
What was you answer?

Light Train

Author:

I created a mind experiment to clarify this situation. It is called Light Train and involves two twin trains on parallel tracks 186,000 miles apart. In one scenario the trains are stopped, and in the other scenario they are moving. The complete setup and diagrams are at the website:

laqu@bravepages.com

Sometimes the site has a password but anyone can get it by e-mailing me at the address on the site. I do not want to go into the complete description of the "Light Train" scenario. It is all listed there at the website. I will say that the ideas discussed there are designed to encourage the reader to study the concepts developed. Also there is a much larger meaning to the setup that I shall let the readers figure out for themselves.

Reporter:
Where did the name Laqu come from?

Author:
It is the name Physlink.com gave me to use. Physlink is a great physics website which anyone can join and discuss physics. It has an estore and a chat room, as well as a place to present dialogue with the other members. The common theme is physics, obviously.

Reporter:
Did you present your ideas there?

Author:
Yes I did. It was enlightening. Not many members talked about "Light Train", but many read the comments. I don't think it was accepted as mainstream physics. One member understood what I was trying to point out but others disagreed with me. Two quite strongly.

Reporter:

So I take it these ideas of yours are rather radical?

Author:

I started with an idea and diagrams and developed them further. They are new ideas perhaps and I haven't read extensively about this type of topic. From my discussion at Physlink I learned that what I am saying, to many people, is considered impossible.

Reporter:

Bucking the current you might say! Sorry, my little joke. But what did you do next?

Author:

I thought about creating a real experiment to clarify my position. Also, I read about Stellar Aberration.

Stellar Aberration

Reporter:

What is Stellar Aberration and what does it have to do with your topic?

Author:

Stellar Aberration is an astronomical concept found by James Bradley in the 1700s. There is a reference at the Iacu website to an article on this topic. In fact the web is an excellent source to study topics I refer to. This topic is very fascinating and the principles I develop can explain the concept of Stellar Aberration. It is another major magic trick of nature.

Reporter:

Tell me more about this and how it ties in with your work.

Author:

That is for another day. Understanding the principles developed in the mind experiment "Light Train" and reading about what Stellar Aberration is will give anyone a start to understanding the process, as well as the connection to what I am doing. As I said, there is a reference to a good article describing what Stellar Aberration is and some of the history of that story. I certainly recommend anyone interested in any word or phrase or concept I discuss to do an Internet search on the topic and learn about the topic for themselves.

Reporter:

Oh, come on! Give me an inkling of what it is.

Author:

Okay, This is how I describe it.

Below is a diagram to help you visualize the earth's vector of motion in relation to a distant star. The sun is the small yellow dot in the center, the earth is the green dot and the vector of motion is the black arrow. The star sending its light would be many miles away on this scale. The vector of motion is a term referring to the magnitude (size) and the direction of the motion.

A vector has size and direction to it. The longer the black arrow is, the faster the motion would be. The direction the arrow points in, is the direction the motion is heading. Take a few minutes to understand this diagram.

The distant star is really distant. About sixteen miles to the right and a bit larger than the yellow dot representing the sun. The vector has no component to the side of a line drawn to the star at A and C. Here the motion is straight to the star, (A) and straight away from the star, (C). At B the motion of the earth is directed at a right angle to the line to the star and the vector has no motion to or away from the distant star.

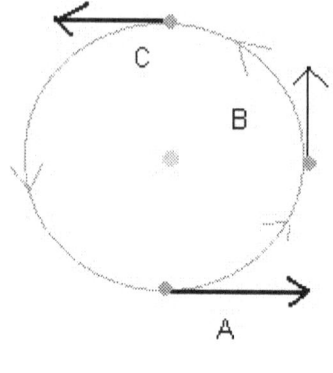

Vectors of Earths in 2-D in realation to a star in the far distance, to the right.
A:motion is to the star
B: Motion is at right angle to the star
C:motion is away from the star.

Figure 1

If one sights in a star with a telescope when the earth in its orbit around the sun has all the velocity directed towards the star being viewed, then that is a certain setting. This is like position A. As the earth moves in its orbit the velocity of the earth, in relation to the star becomes perpendicular to a line drawn to the star and earth. This is three months after the initial sighting, when the velocity was zero and is represented by position B. Obviously the earth's velocity is not zero but the velocity directed towards the star can be. Now at this point, the setting on the telescope has to be changed to see the star. Guess whether one has to look ahead of the known position of the star or behind the known position.

Reporter:

I would say one has to look behind where the light is coming from, since it seems we are moving past the star.

Author:

No, you have to sight the telescope ahead of where the star actually is. And the picture you took of the star when the velocity was zero, compared to the picture when the velocity of the earth is at maximum, shows the star behind the first position. One has to look ahead to see a star because of motion. And if you take its picture, the star will be behind the picture spot when the motion to the star was at its maximum. Also, the position in the telescope moves back to the starting position when the Earth gets to position C. Here again, all the motion of the Earth is directed away from the star and there is no motion at right angles to the star.

Reporter:

Are you sure of this? I need another break.

Author:

That will give you an incentive to read and study this topic, to understand what is going on and how nature does this trick.

Later that day.

Reporter:

Before we go on could you tell us the name of the instrument you built?

Author:

Certainly, I'd be glad to. It is called a Velador. This stands for "absolute velocity detection by optical radiation". Pronounced like "VELLA DOOR". I use it as a generic term such as telescope or microscope. I assume there are going to be many variations and modifications of this basic design. The current one I am using is the third design and I have another design, which I am currently working to build. The first of any prototype is usually very simple such as the first telescope or microscope. Now look at how those instruments have evolved.

The fourth variation will include a rotating horizontal platform holding supports straight vertically that will suspend the 10-foot pipe with camera and laser. This will allow 360 degree positioning horizontally and as well as to allow the pipe to be moved 360 degrees in any setting. This will be used to hopefully pinpoint the position that gives no motion to the laser beam as measured by the camera.

Hands On Experiment

So I sat down and tried to think of an experiment that could demonstrate my point that motion influences what we see. Specifically, an experiment that would demonstrate that light could seemingly turn under a field of motion. I needed something simple that I could manage. After all, I am not a university or large publicly funded research group. The first thought I had was to try and make a situation resembling the "Light Train" experiment.

I came up with an idea for an experiment using a fixed distance rod, which was a ten-foot steel pipe. At one end I tightly fastened a laser. At the other end of the pipe I firmly affixed a CCD (charged coupled device) camera. A CCD camera has a special plate made up of many small light detectors called pixels. If the camera has two million of these detectors, the chip is a two-megapixel chip. If there are five million light sensors, then it is a five-megapixel chip.

The rod, with the camera and laser attached, is mounted horizontally on a turntable or wheel. The wheel has a smooth bearing, to minimize roughness as it turns. The wheel that the rod is mounted on spins the rod parallel to the ground.

If the earth were completely stationary in space, my reasoning was that the laser beam would strike the CCD at the exact same point when the rod was rotated. If the earth were in motion, then the time it took the beam of light to reach the CCD would be ten nanoseconds. Light moves about one foot in one nanosecond and a nanosecond is one billionth of a second. This would give the CCD time to move in the field of motion of the earth. With just a little motion in that ten nanoseconds, then the laser light beam would not hit the same spot it would if the earth were still. And, if in fact the earth were moving at different velocities, the beam would then hit different areas of the CCD.

Obviously the earth is moving. An example of how one velocity vector can seem to be different is the analogy of a car driving 60 miles per hour (mph) due north toward a mountain. Imagine a passenger with a powerful range finder that can measure speed accurately. When the velocity is measured due north, it is sixty mph due north. Remember, the speed of the car is always sixty mph but the velocity is a combination of the speed and the direction. Then the speed finder is moved to the east. Here it sees no motion, so the velocity of the car is zero to the east. It would be a minus sixty mph to the south and again zero to the west. This is perhaps a bit confusing. How fast is the car approaching the mountains? Sixty mph. How fast is it approaching something to the east? Zero mph.

Reporter:
How did this follow from the Light Train concept?

Author:
I imagined the CCD replaced one of the trains and the laser replaced the other. The steel pipe kept them an equal distance apart and parallel to one another. Now the motion of the earth would act to move each piece at the same

velocity. This would allow me to study the effect of changing direction and velocity on the light ray striking the CCD.

Reporter:

Why doesn't the laser beam follow the motion of the source and hit the CCD on the exact same spot, whether it's in motion or still?

Author:

That is the beauty of a beam of light. Once released, light does not know what lies ahead or behind it. All it does is move straight ahead. It has no motion or at best very little, to the side, up or down. It behaves just the same whether it is sent from a stopped source or a moving source. It behaves, as if all contact with the source has been cut off and the effects it has is on whatever it interacts with. The detector is what effects it. Motion, form, color, change of position, all these characteristics and more depend on the detector. I have read there is actually a small spread of a light ray to the side, up and down but that effect is very small and not enough to account for any major change.

Reporter:

OK, but the beam in your set up will always hit the same spot on the detector. You can't know if it were meant to hit ahead or behind the struck spot in the case of a motion field or a still situation.

Author:

That's true. Take a very simple situation. Say the beam and detector are arranged due north and south with the beam shooting to the north. We know the earth is rotating to the east (This motion produces day and night and the rise and setting of the stars and moon) and just in this case let's consider the eastward motion only. We record the spot where the beam strikes the CCD, then turn the apparatus so the beam faces south. The detector is now facing south and the side facing east previously is now facing west. The detector is riding in the field of motion but now in a different direction. The light will hit at a different spot if this is true. In other words, when the detector was facing north, it was moving with its right side moving east, but when it is turned to the south, the left side is now the side moving to the east. The beam always strikes behind the detected spot, from the direction of motion. Thus if the detector is moving and a laser beam is shot at it from ten feet away (roughly perpendicular to the face of the detector), the light should strike the detector behind the spot it would have if the detector were not moving. After all, light does not know if the detector is moving or not. In fact, it does not know if a detector is even in the path it will take. I set up such an apparatus and have a series of pictures, which I will show you and attempt to explain.

Eye Of the Tiger

First I created my apparatus. I had a wheel, placed on its side, mounted to a small table. I had it mounted horizontally so that the wheel could spin smoothly in a full 360 degree circle. Then I attached a ten-foot pipe to the wheel with bearings. The center of the pipe is attached to the center of the wheel with a strap, so that it will balance and not fall off as the wheel is turned. I painted the pipe black to minimize any light reflection. The pipe is attached in the center of the wheel so that the ten-foot pipe extends out to the side of the wheel by five feet on both sides. Now I have a pipe that I can spin in a smooth circle.

To one end of the pipe I securely attached a laser light and to the other end I attached a camera with CCD. The laser is pointed directly down the pipe at the CCD.

In this series of pictures, the camera and laser are mounted on the inside of the pipe, which is four inches in diameter. Later, I attached the camera and laser to the outside, as it is much easier to do that way. I took an initial picture, then rotated the pipe containing the laser and camera thirty degrees, and took another picture. I continued to rotate the pipe (and chose to rotate it about thirty degrees each time thereafter) and take pictures. Here are my first series of pictures. And I have to thank you for making these available in color on the separate cover display.

Series A:

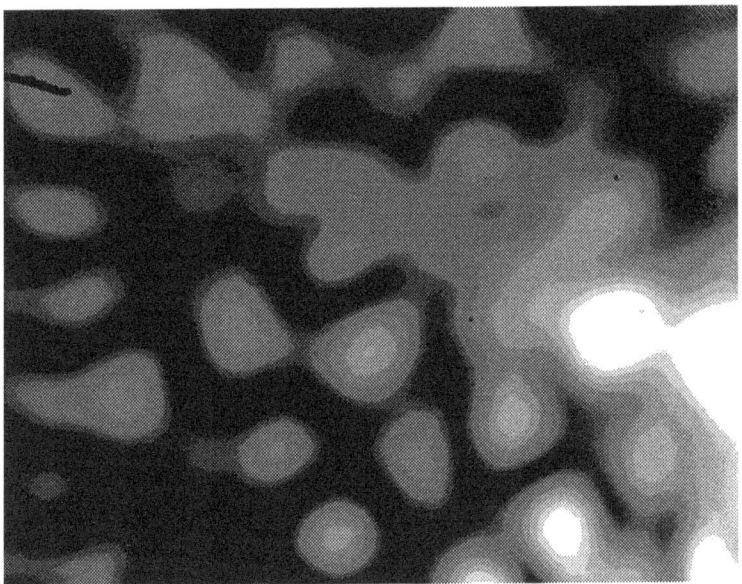

Series A 1

This is the first picture I took. It is a beautiful picture of an interference pattern generated by the red laser. I believe the laser ray produces an interference pattern, which the CCD records as the red circles and black spaces. The black straight object in the upper left is a piece of dirt on the CCD. I believe the purple areas in the lower right are due to a more intense light hitting the detector.

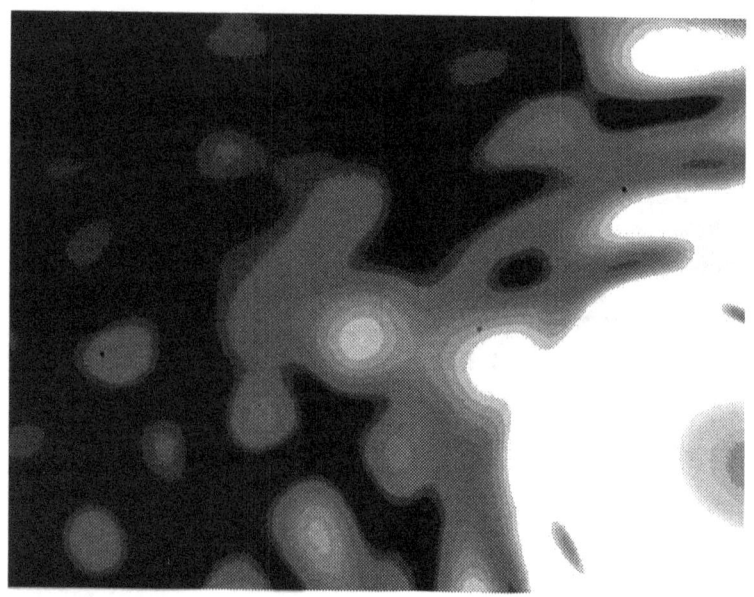

Series A 2

Next, I rotated the pipe about thirty degrees, and took this picture. At first I didn't know what the blue image was at the lower right of this picture. Now I assume it is the laser's direct image. Blue perhaps because the intense light from the red laser washed the red out, leaving the color blue to denote where the laser was shining. Anyway, I wondered about how the center of the laser suddenly became visible. I hadn't changed the relationship between the laser and the CCD, but instead had rotated the apparatus on the wheel thirty degrees. To my thinking, the laser and detector are in the same position they had been in the first picture. The only difference is that the entire apparatus had been rotated thirty degrees.

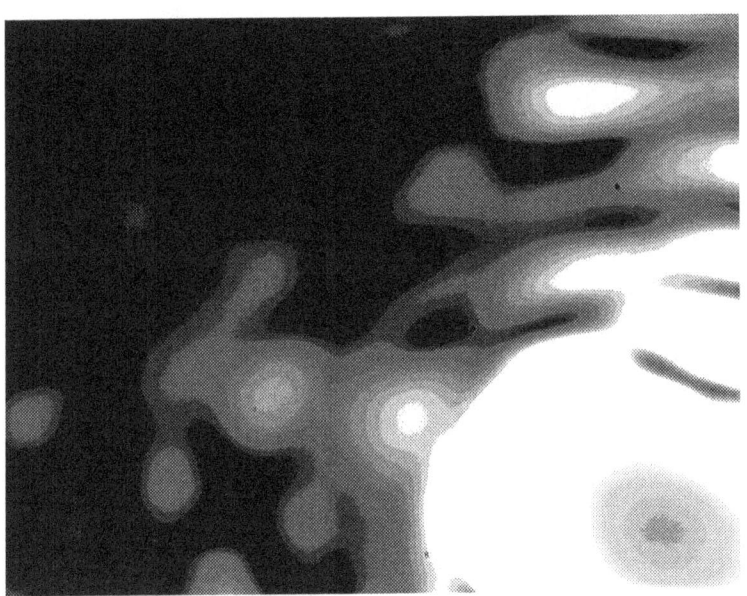

Series A 3

Again, I rotated the pipe by moving the wheel another thirty degrees. Remember, neither the camera, nor the laser has been shifted in relation to one another. The beam captured on the CCD has moved farther to the left. We can see the total beam impact area. Dark blue, then lighter blues, then even lighter blue, then a white area. All this just came from rotating the pipe on the wheel.

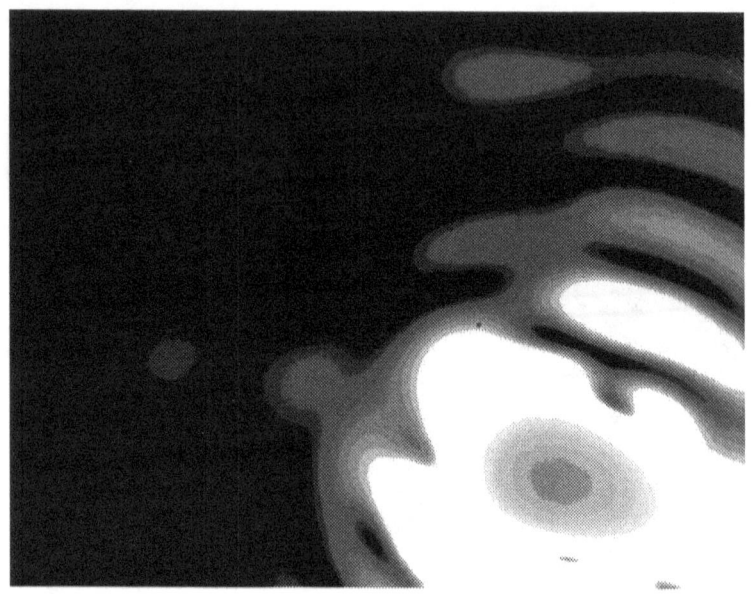

Series A 4

In this picture, I moved the wheel so that the pipe is rotated even farther. The spot again moves to the left. Each picture represents an image with the CCD and laser in the same position with respect to one another. However, as the pipe is moved to different directions, both the camera and laser are in different fields of motion from the earth's vector of motion.

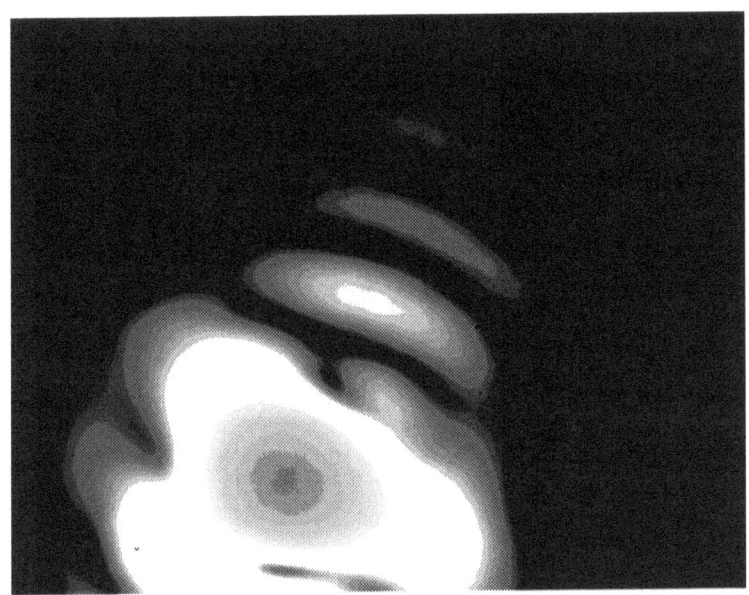

Series A 5

For this picture, the apparatus is again turned and the spot continues to move left. Interference rings are still visible. When I saw this laser image I was very gratified and excited. Here was a primitive bit of Nature caught on a CCD image. The spot has energy. A still picture shows it quiet, but looking at it in the camera's image display showed it to be a shimmering structure. The colors are artistically satisfying. I am sure there is a lot of science happening in just this limited picture series. The generation of the laser spot, the colors, and the interference patterns are just a few of the marvels.

This was one of the first series of data that I recorded and I call it "The Eye of the Tiger". I had been perusing an elusive beast for several years. This was the first time I had seen what I was after. It was an exciting, stunning and never to be forgotten moment. I was overwhelmed with elation when I saw the brilliant laser spot encoded in the bright blue and black colors move across the CCD and return to its starting position as the pipe was turned through the 360 degrees.

Author:

Now let me say a bit more about the experiment I created.

Reporter:

Excuse me, but do you have a visual of this apparatus you are using to record these images?

Author:

Sure. There are pictures in the back of this book of the setup.

Reporter:

Yes that helps, except the pictures are somewhat blurry.

Author:

Sorry.   They are the best I have of that set-up and the pictures did not copy well.

Reporter:

Shall we continue or take a break?

Author:

Break time.

Later.

The Experiment

Author:

With the encouraging results of the test run, which indicated some motion of the light beam, I made the following experimental set up.

Reporter:

Before you describe the actual experiment, would you clarify how you measured the light beam on the CCD? How could you tell it was moving?

Author:

Using the CCD, I recorded the image of the laser light beam at the various positions the wheel was turned to. I rotated the wheel 360 degrees, stopping at fifteen-degree intervals. Thus there were twenty-four pictures. Later, I stopped at thirty and forty-five degree angles to save time. I then analyzed each picture. I did this by loading the pictures onto my computer and saving the image as a Jpeg image, or a bit map image. I then saw the lasers blue spot move across the picture and then come back.

Each image has a viewing program to put the picture on a computer monitor screen. Each program has a cursor that you can move around on the picture and as the cursor moves a set of co-ordinates is generated by the program. I put the cursor on the center of the laser image and recorded the pixel co-ordinates that the program generated.

Reporter: Fine, and you then could use the numbers to mark the laser center for data to analyze.

Author:
Yes, that is correct. I plotted graphs of the movement and distance of movement. I figured out how fast the detector would have to move to account for the position shift.

The results of these pictures were fairly impressive to me. I had a thought after reading a book, I devised a thought experiment, studied Stellar Aberration, then designed an experiment to justify my ideas. This was not a random pattern of events.

Ready for the complete description of the experiment?

Reporter:
Yes, go ahead.

Author:
Color versions of these pictures in Series B can be found in color on the cover. Also the previous pictures in "Eye of the Tiger" can be seen there in color.

Experiment
To
Measure The Absolute Motion
Of the Earth

By
Lance Osadchey
04/22/2003

Hypothesis

Since a single light ray is believed to exist in its own space-time and proceed straight ahead in propagation at the speed of light and a laser ray approximates a single light ray, these rays have no (or minimal) lateral or sideways motion. Thus it should be possible to measure the absolute motion of the Earth (or anything). One has to be able to observe the position of a light ray on the object moving and know the distance from the source of light to the object and be able to figure the change of the impact of the ray of light in different views.

Equipment

A solid, minimal-vibration table or optical bench

A solid rotating bearing

A steel pipe ten feet in length

A laser source with positional holder

A series of lenses to retard the laser beams intensity

An optional series of lenses to direct and focus the laser beam

A CCD for detection of the laser beam, with capability to record the image of the laser impact and to record a series or a movie of the laser spot's motion upon the detector

Suitable equipment to analyze the laser spots position on the image from the CCD

Procedure

1. Use the table as a stationary support for the apparatus.
2. Attach the bearing in the form of a rotating platform to the table.
3. Attach the steel pipe to the bearing with the center of the pipe at the center of the bearing (wheel).
4. Use sorbofane to dampen vibrations under the table's legs and under the steel pipe.
5. Attach the CCD securely to one end of the outside of the pipe.
6. Attach the laser with the positional holder to the other end of the pipe, on the outside.
7. Attach the lens to attenuate the intensity of the laser in-between the CCD and laser but close to the laser.
8. Record the image of the laser on the CCD at positions one (for example, due north).
9. Rotate the pipe around its midpoint by turning the wheel in a circular fashion, at various increments and record

each image by taking a picture at each station for a complete revolution.

10. Determine the position of the laser's center numerically on the CCD and then compare the positions recorded at the various stations.

11. Make measurements at fixed places on a circle of rotation.

12. Use a measurement system for analysis (such as bit map) to locate the laser center and plot the motion of the impact site. Use the center of the laser spot for the point of measurement.

Results

I used a bitmap image program to analyze my data. Creating a bitmap allows me to place the pointer on a certain spot of the picture and there is a read out of the x and y co-ordinates of the spot. It is difficult to judge the exact center of the laser image but with practice one can come within five pixels of each attempt.

Some runs, where there was no turning of the pipe, often showed a drift of a constant quantity, depending on the position of the apparatus. I did several standard non-turning runs as well for comparison. These runs often showed no motion of the spot on the CCD. The difference between the runs where the pipe was turned versus the runs when the pipe was held in one place was thought to be a function of the time taken for the turning, which were one to two minutes. Where as the drift patterns from static runs were of several hours.

Here are representative images of a typical run. Each station is approximately thirty degrees apart. A typical run begins at zero degrees and moves counter-clockwise through the stations, finishing back at north (zero degrees) after a complete rotation.

This particular run shows the laser beam to have a black center. This is very interesting since black would seem to indicate no color at this area. I am not sure if the black color is a function of the CCD getting too much light, but I think not, as that would produce a white spot. The black color may indicate a property of light from a laser that masks out the color at the center. Any way, to me it is a beautiful picture.

I took these pictures on March 23, 2003. In some of the images, the center spots were actually off the picture. However, it was an early run and because of sentiment it is the one I chose to include. This was one of those fine days, full of excitement and awe.

Series B:

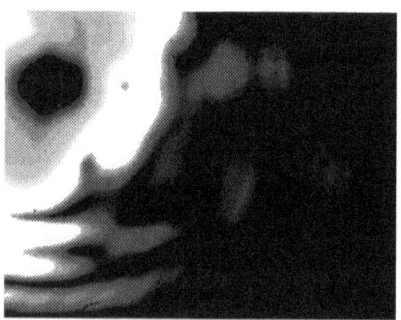

Series B 1

This was the beginning station, about due north. That is, the laser was shining from the south to north, with the CCD at the north end of the pipe facing to the south.

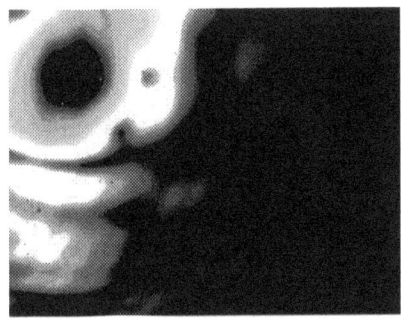

Series B 2

In this picture, I moved the pipe counterclockwise thirty degrees. Again, the camera and laser were ten feet apart, giving the light beam ten nanoseconds to cross this distance.

Series B 3

This picture was taken after I rotated the pipe sixty degrees.

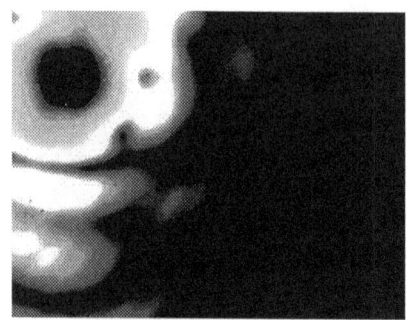

Series B 4

This shot was due east, rotated 90 degrees. Directions of the compass are good for measuring where you move on the earth's surface. But for telling what is happening in the universe they are inadequate. The earth's vector of motion, using the stars as guide posts would be interesting and more informative.

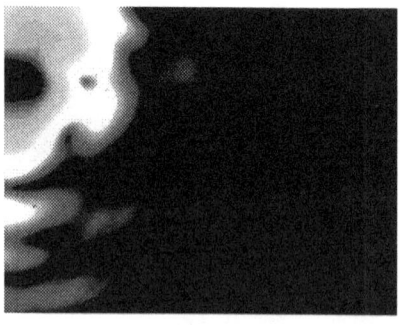

Series B 5

This is 120 degrees from the start (past east heading south). Note that the dark center of the spot is moving off the edge of the picture.

Series B 6

This picture was taken after the pipe had been rotated 150 degrees from the starting position. This would be almost due south. I never got a smooth transition from point to point, either because of the crudeness of the instrument or due to the vector constantly changing its qualities. Possibly it was a combination effect.

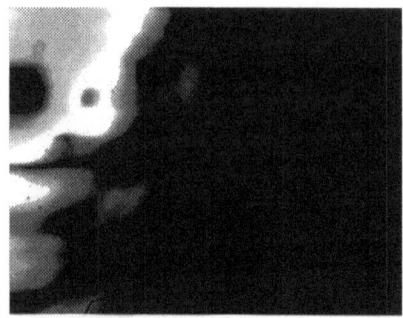

Series B 7

Due south is the compass position. The pipe rotated 180 degrees.

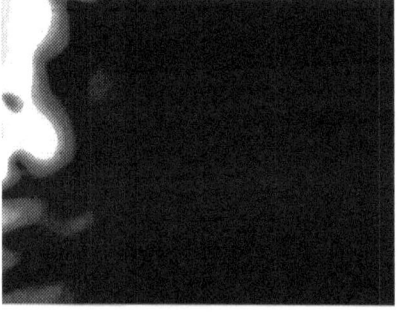

Series B 8

210 degrees, now moving to the west.

Series B 9

This picture shows the image after rotating the pipe 240 degrees from due north.

Series B 10

Due west is this position, or 270 degrees from the beginning position.

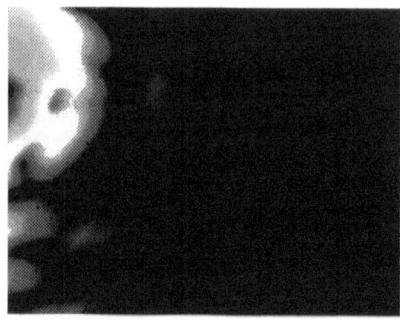

Series B 11

This is 30 degrees past west in a northerly direction, or 300 degrees from the start.

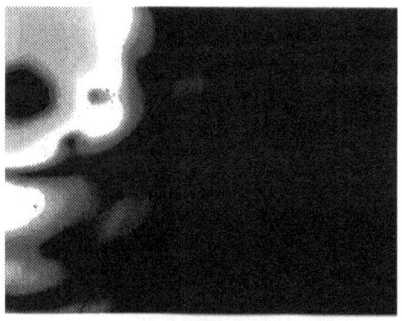

Series B 12

30 degrees to north approaching from the west, which is the same as 330 degrees from the initial position due north.

Series B 13

Due north again, after a full circular rotation. Compare this picture to B1 and you will note the laser center is very close in orientation. It is not exactly the same possibly secondary to the accuracy of the instrument or possibly the vector has moved from the original setting to a new position.

Considerations

The pixel size I used was considered to be 4.2 micron (one micron is one part of a thousand of a millimeter). I e-mailed, called, and faxed the company that produces the camera for more information about the pixel layout and size, as well as whether there is a program inside the camera that maps the image on the CCD left to right and up or down. Unfortunately the company had no information for the public at the time of their reply. They did acknowledge that the camera reverses the image, but did not say if there is a mapping of the CCD to the view-site that reverses the image. Obviously if the pixels are, for example, eight micron in size, that makes a difference in the calculations of the velocity. The technical manual that came with the camera indicated the pixel size was 4.2 microns, so that is the number I used.

However, believe it or not, those are technical details that can be worked out later and the actual numbers are not as important to me as the principles that are being developed. At present, the instrument is not as sensitive as I wish it was and the degree of accuracy is not as sharp as it could be. However, this is just the beginning and I hope that later refinements and improvements on the equipment can be made.

In this series, the measurements were approximately as follows. A displacement of 120 of the pixels on these pictures represents about 300 pixels on a 1200 by 1600 bitmap. In this series, low-resolution images were taken. Multiplying by a factor of two and a half did convert these to standard size. This gives a displacement of about 300 pixels. Displacement is another term for distance of movement. Since each pixel measures 4.2 micron, this then gives a displacement of 1260 micron. Again there are 1000 micron in one millimeter. This distance, 1260 micron, is half the total displacement. Since the laser spot moves in a curve on the photographs it is possible to calculate a center position of the curve. The center is considered the position that would be struck by the light ray if there were no movement of the CCD.   Because the earth is moving and the CCD is carried along with it, any deviation from the center is considered to be caused by the movement from the earth's_motion. Thus, if the distance the ray traveled from the laser to the CCD is about ten feet and light has a velocity of about one foot per nanosecond, then the motion of the CCD must have been about 126 kilometers per second, to account for the displacement. I am using half the value of the total displacement since, as the CCD is turned, it moves with the same velocity but in an opposite direction. In other words, if the maximum displacement is exactly north, then in the south direction the velocity will be about the same except in the opposite direction. This is just that particular view of the vector at that day and positioning of the CCD. The various force vectors the CCD is subject to would be constantly changing. My belief is that by the time one has recorded the data and gotten back to the starting direction, the appearance of the vector has changed.

The laser impact spot measured on the bitmap to be 108 pixels, so that the actual size was 1134 microns or 1.1 millimeters. The actual bit map was 1200 by 1600 pixels and the company said that each pixel was mapped to each separate bit. This would make the CCD size to be 6720 by 5040 microns or .6 centimeters by .5 centimeters.

On some runs the small black center measured to be 60 pixels, which corresponded to .252 of a millimeter. These runs provided a better assessment of the actual movement of the spot.

Conclusions

The CCD camera was attached firmly to one end of the pipe. The laser was securely attached to the other end of the pipe. The pipe was rotated on the wheel bearing in a ten-foot circle, stopping at various points on the circle and taking a picture. This produced motion of the impact site of the laser light on the CCD. I conclude that the CCD was moving at different velocities at each station where the picture was recorded, in relation to the light beam.

The velocity of the CCD picked up the velocity of the earth. Since the pipe was ten feet in length, this gave a ten-nanosecond time from initiation of light at the source, to detection at the CCD. During this time, the detector moved slightly. As the detector is moving with the velocity of the earth, it recorded the velocity at various angles as it was turned. The light ray, once released, carries no component of motion of the source. Possibly because initiation of light is so much faster than the motion of the source and/or light carries no memory of the sources motion or a photon just goes straight. It is believed photons have no mass and this could lead to the conclusion that they do not pick up sideways motion.

I did trials with the wheel revolving and with the wheel stationary. In the stationary trial, the impact area on the CCD was the same. The impact area on the revolving trial showed a repetitive motion of the impact site, in an approximate ellipse.

Acknowledgements

I have carried this experiment about as far as a single person with limited money for equipment can. The set up becomes difficult using longer pipes, as the holding and housing of a larger pipe is difficult.

Here are some improvements I'd like to see. Obviously a better CCD, with more pixels per unit area, would be instructive. A secure bearing and better support for the pipe is essential. A better laser would be helpful, with a better laser sighting system. Using simultaneous setups with laser and detectors compiled to an analytic program would also be useful.

It would be beneficial to make a three dimensional plot of the vectors as a function of the plane the measuring system is in. Vertical-horizontal and left-right planes are needed and should be correlated. I visualize a three dimensional plot showing the major velocities as rounded protrusions from the center of the plot, never actually touching the exact center.

It would be instructive to have a set up so that the pipe could be rotated in a vertical fashion, as well as the horizontal rotation that this work described. Having the recording done by a computer would be helpful. Two simultaneous set ups would be useful for comparison. A computer program would analyze the data better as well as being able to link the data to celestial coordinates instead of earthly directions.

Once mapped, the vector and how the position of the vector changes with time could be mapped to the celestial co-ordinates. This could determine the relationship of that motion as it would appear mapped to co-ordinates using the universe as the overall picture.

Future Work

Refinement of the experiment is needed. Better equipment and a more stable setup, with better computer analysis is required. How this and the results fit into present day theory remain to be evaluated. It would be exciting to see a three dimensional model of the vector of motion of the earth. It is possible to do this now that the groundwork has been preformed.

I set out with this experiment to prove what I believe has been shown. Work on Stellar Aberration is consistent with this result. I created a thought experiment entitled Light Train to explore some of the concepts that led to this experiment.

Reporter:

Well, I have to think about this for a while. Could you go over your procedure from the data of the equipment to defining the vector of motion?

Author:
Yes. As you know, one of our systems of defining the world rests in having three dimensions to use for the location of items in space. Up and down, left and right and front and back.

If we want to use a graph, we use a three co-ordinate system. A system that has an x-axis, a y-axis and a z-axis. This three-dimensional system has a host of properties associated with it. A point has no dimensions. A line has just length. A plane surface has two axes and describes a flat structure with only two dimensions. Here lines, circles, squares, rectangles, triangles and the like exist.

Next are the three dimensional objects existing in a three dimensional space. Spheres, boxes, pyramids, cubes and solid objects exist here. Now if I want to measure the length of a diagonal in a box, a line from one corner to the further corner, how do I do it? Imagine a box with one side being two units, another side being three units and the third side having four units in length. A unit is a way of saying you can pick what ever you want, centimeters, inches, feet, or yards. All are acceptable.

An easy way to see how the measurements are done is to first figure out the length of the diagonal of the side that has two and three as lengths. This measurement is found by figuring the sum of the squares of the sides, then taking the square root of that. Remember the Pythagorean Theorem for a right triangle?

So, if we add the squares of two and three we get thirteen. Now we take the square of that and get the square root of thirteen. That is the length of the diagonal of the two sides we are working on. Then, if we rotate our box so that the diagonal is pointed vertically and line it up the other side, which was four units, we see another right triangle. This triangle has a side measuring the square root of thirteen and the other side measuring four units. We then use the sum of the squares of these sides and take the square root of this to find the diagonal. This is the square of the square root of thirteen (which is thirteen) and four squared (which is sixteen), giving us twenty-nine. Taking the square root of this now equals the square root of twenty-nine, which is the final length of the long diagonal of the box.

The formula to do this, a bit easier, is the square root of the sum of the squares of the sides. If the sides were a, b, and c the formula is equal to the square root of a squared plus b squared plus c squared. There is a picture at the end of this book which shows the long diagonal in a box with sides, a, b, and c.

Now that we can do that, the next question is: how many planes do we need to define our vector?

Reporter:
Well, as I see it, one plane can only define two dimensions of the vector. So one would need another plane to do it. So it seems two planes would be enough.

Author:
Yes, that is the way I see it also. However, one has to be careful in choosing the different planes so they are comparable and give stable results. Imagine our three

planes are all at right angles to one another. Define the plane straight ahead as having a y-axis that runs vertically from point zero. Also it has an x-axis running horizontally starting at point zero. The y-axis has numbers, which increase as you go up and the numbers on the x-axis increases as you go to the right. If one looks at a three-dimensional object using this plane, all that shows up is a flat object. It has height and width but no depth since it shows only two dimensions.

In my experiment, I took measurements at one of these planes. I had to define which direction I was pointing the CCD and use that as an orientation plane. I then turned the wheel 180 degrees so it faced backwards and the y-axis still measured from zero vertically, but the x-axis is now backwards from the starting position and is measuring increases in the opposite direction.

Reporter:
Explain that please.

Author:

Let's say the x-axis in its first position was measuring from zero to the right as it increased in value. That direction would be east if we were facing north at first. Now we turn and face south. The x-axis is now measuring higher values to the west. Note that the y-axis is measuring up, as higher numbers in both cases. The z-axis is going from front to back in both instances and is not measured. I used this changing one axis position, while not affecting the other two dimensions so that I could calculate where the laser dot would strike if there was no motion to the system. It would be for example, if x equaled plus ten on that axis facing north and plus fifty when facing south, I said the expected value, if there were NO motion, would be half the value of the two points. In other words, the dot without motion would be in the middle of the two points. So in this case with motion the values were ten and fifty. Half the difference would be twenty points and this added to the ten gives thirty. With motion, the values are deflected twenty points from this middle either way. If there were no motion of the detector, the beam should strike in the same spot each direction, namely at thirty.

Reporter:

All right so far, any more on this topic?

Author:

Yes, now that the x point has been found, we have to do the y and z-axes. By carefully selecting and turning the apparatus, each axis can be isolated from the other two and the center point of each can be found. Take the y-axis. By rotating the wheel 180 degrees, then turning the detector by 180 degrees, the z and x-axis are in a similar configuration to what they were at for the first measurement. Yet the y-axis is "upside down" and measuring in a different direction. I did not do the three dimensional measurements, but instead stayed with the two dimensions. Obviously, three dimensions are needed to truly define the vector.

The z-axis can be picked up from the starting position by rotating the wheel just 90 degrees. The y-axis is the same, but now the x-axis has been situated so that it is in the front and back of the detector with the z-axis running in the x-axis's old spot. By isolating the different axes, then turning the apparatus just right, one can get the values for x, y and z. Using the equation to solve for a diagonal, the value of the vector can be calculated.

The conversion from pixel to micron has to be done also. In my set up, I had only one laser and detector, thus it took time to change the position. I suppose the orientation of the motion could change during that time. Ideally, a three laser system with simultaneous measurements would be very helpful. Also a computer to keep track of the measurements and do the calculations of where and how big the vector is.

Reporter:

But then all you have is a distance and no idea of the time element in the velocity.

Author:

Just use the ten nanoseconds the light took to get from the laser to the CCD and that is the amount of time it took the CCD to move, giving the shift of the laser spot. The CCD is moving in the field of motion of the earth. If the calculations show a CCD movement of 280, then to go from that to kilometers per second all one has to do is move the decimal one place to the left. In this case it would turn into 28 kilometers per second. Then to get miles per second multiply 28 by .62, which would equal 17.4 miles-per-second. If you wanted to see the result in miles per hour, multiply the number of seconds in an hour by 3600.

Reporter:

How can you divide micron per nanoseconds by ten and get kilometers per second?

Author:

Take a piece of paper and pencil and figure that out for yourself. It may be more convincing that way. Here is a paper with the rationale on it. Put this as an example in the Work Area at the end of our conversation.

Reporter:

Since we have some time left, could you tell me your most gratifying moment of this experiment?

Afterthoughts

Author:

Let me say there have been many moments of satisfaction. Each step was a different experience for me. Since it was all new ground, I had to think out the steps to take and the procedure. I have to understand something myself, not just read about it factually and accept it for the words or formula. I want to know how things work on a fundamental level, as much as possible.

The most astounding moment came when I directed the laser into the camera and recorded the spot of light. I found that by moving the pipe setup, the laser image shifted on the CCD. First, it was a beautiful picture of the interference pattern and the solitary laser beam. The colors were very vivid. Then to have thought that the image could move with the turning of the wheel and to see that I could measure the motion of the spot on the CCD was both gratifying and astounding.

It was as if, by visualizing the motion of the brightly colored laser image upon the CCD, my journey involving tracking and hunting for this elusive animal was nearing completion. I knew the idea would work. Although the elusive prey was not captured (and will never be totally captured), I had seen the tiger for the first time.

Reporter:
Good. Any other memorable moments?

Author:
Three months ago it was as if I was looking for an event with a fine-toothed comb and magnifying glass. On one set up, I had the laser shining directly on a large piece of cardboard. I thought that by turning the whole apparatus around the inside center of the pipe (not on the wheel but just turning the pipe) with the cardboard where the camera was and with the laser still attached, it might show something. To my amazement, the laser's bright image, as the cardboard

was rotated with the laser fixed on it, described a circle where the laser shone. Unfortunately, it was a large circle. I thought this meant that the motion of the earth transmitted through the cardboard so that instead of the camera being a detector, the cardboard itself was now a detector. At the time, I thought the motion was so large I didn't require a camera for fine movements but a simple section of cardboard would pick up the motion. When the cardboard was attached to the wall of the house, it produced an even larger circle, as one would expect. Doing this resulted in two circles on the cardboard. One circle was large which was from the cardboard being rotated and one circle was small, which was from just rotating the laser. Interestingly enough, the little circle always intersected the larger circle at two points. At the time, it seemed as though Mother Nature was running through a large house while I had been looking for her with delicate instruments. It was as if I turned a corner and there She was. She hit me in the face with a pie and ran down the hall laughing. However, as I studied this more, I became to believe it was a bend in the pipe, as I had attached another ten foot section of pipe to the first ten foot section. Despite tightening the joint as much as possible, I suspect that it still allowed for some extra motion and this produced the smaller circle. I have not had the time to study this effect more, but wouldn't it be strange if there was actually a small circle there, and not the large one created supposedly by the bend in the pipe? That's what anyone is up against when you study new phenomena. It would be then like getting two pies in the face!

Reporter:

Let me ask you, why aren't there more formula in your work?

Author:

At this point I have no detailed formulas to apply to this work. The instrument has to be redesigned with a better CCD and laser. Once accurate data is acquired, formulas can then be developed. It would be a waste of time to devise formulas now with what little data I have. I have settled for the principle of the idea. This is the skeleton of the project and later formulas can be calculated and devised.

Odds And Ends

Reporter:

Do you have any further comments?

Author:

Yes. I would like to say a few words about the concept of the vector of motion of the Earth. I picture the Earth rotating on its axis every twenty-four hours. It also orbits around the sun in a year. (One run of pictures I took suggested a periodic orbiting of about twelve degrees of change a day. This may imply the rotation of the Earth around the common center of rotation of the moon and Earth.) The sun is also moving through space in our Milky Way galaxy. I envision all of these motions combining to form a smooth helical pattern as the Earth moves in space. Viewing this motion from different angles and directions can produce various patterns of motion that range from simple to very complex. This is where computer analysis of data and creation of the vector from the data would be very instructive.

Also a few words about "Light Train", Stellar Aberration and this experiment: in a way, despite what I have explained, there are certain factors to all these topics I have left out. I want people to think about the topics and come up with their own conclusions. There is a common group of threads running through all three of these topics. I have not correlated them. I believe you should figure them out by yourself, as this is much more instructive then me just telling you.

Reporter:
All right then, give me a hint or two to be sure you are not just blowing smoke in the air by claiming that.

Author:
I really like how direct you are. For the thought experiment "Light Train", there is a correlation to something much larger, or I should say smaller in this case. Take the principle of the thought experiment and apply it to something else.

The thought experiment is straightforward. However, it is based on something else. Something that exists now, and something from the Victorian Era.

Try and identify the common theme of the three descriptions.

The topic of Stellar Aberration also contains principles of Nature, so nothing can be done to manipulate these. One has to think about what Nature is saying and, from a very limited personal view, explain what one sees. It is much like the three blind men describing an elephant; a lot depends on where you stand.

Now let me ask you, as we are close to the end of the interview, who do you think turns the light?

Reporter:
I am embarrassed to say I am not sure. I have to think about all of this.

Author:

Well in that case, let me provide you another clue to understand this process. Imagine that light moves instantaneously. As soon as the laser is turned on, it hits the CCD at a certain spot (and will always hit this spot unless the laser or the CCD is moved). However light does not move that fast and there is a bit of a lag before it gets to the CCD. This provides a bit of time for the CCD to move, if it is moving. That is the key principle.

Work Area

I have tried to explain definitions and scientific topics as I went along. The Internet and World Wide Web are tremendous sources for additional information. This is one reason there is no formal bibliography.

Calculations on the River Crossing Problem

Imagine a river beginning at the left side of the page and flowing across the page to the right. It has two courses set up on it. Assume the course for swimmer A to swim across the river is 2.5 miles long. The other course is for swimmer B to swim down the river 2.5 miles. Both swimmers can swim 5 miles per hour (mph).

In the situation where the river has no current both A and B swim 2.5 miles to the turning point, then 2.5 miles back. Each has swum 5 miles at a speed of 5 mph. It takes them each one hour.

Now give the river a down-stream current of 3 mph. Swimmer A can swim at 5 mph but the current is pushing him to the right. He therefor angles his swimming into the current. This creates a right angle triangle. The current is flowing 3 mph and the swimmer is swimming 5 mph across it, slightly angled. The speed straight across, is the other side of the triangle. By using the Pythagorean Theorem one can calculate the other side, which is the swimmer's actual speed across the river.

So swimmer A's speed is 5 and the current is 3. Plug these into the equation A squared plus B squared equals C squared.

Thus his actual effective speed across the river: 9 + actual speed squared =25. And 16 has 4 as its square root. This makes sense as A can swim 5 mph in still water and with a side current can only go 4 mph in a line across to a point exactly opposite the start. A is still going at 5 mph, just not straight across.

Here is a diagram of the above calculations. Don't forget the vector diagram is not the actual representation of the way the swimmer moves. "A" swims straight across at the 4 mph despite being angled slightly upstream to account for the current.

5 squared = 4 squared + 3 squared

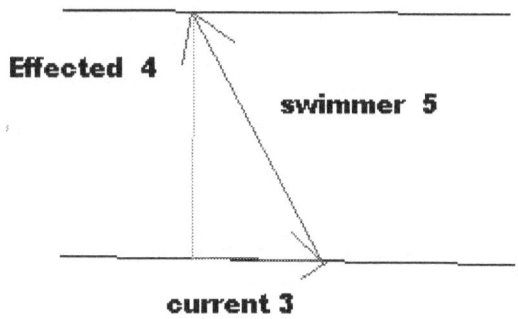

So the time it takes A to do the 5 miles round trip at 4 mph is 1.25 hours.

Now consider the Swimmer B. Going down current, the swimmer's speed is 5 mph plus the current of 3 mph, to equal 8 mph. So the swimmer swims 2.5 miles down at 8 mph. Coming back, B's effective speed is 5 mph minus 3mph, which equals 2 mph. Swimmer B's time back upstream would be 2.5 miles at 2 mph. The total time it would take would be .3125 hours, going with the current, plus 1.25 hours, coming back against the current, for a total of 1.5625 hours.

In fact, for all currents up to the speed A and B can swim, A can do the course faster than B. If the current is 5 mph, or faster, Swimmer A can never get across to the opposite marker and B will never be able to get back.

Long Diagonal

Picture of a box with sides 2, 3, and 4 with the long diagonal:

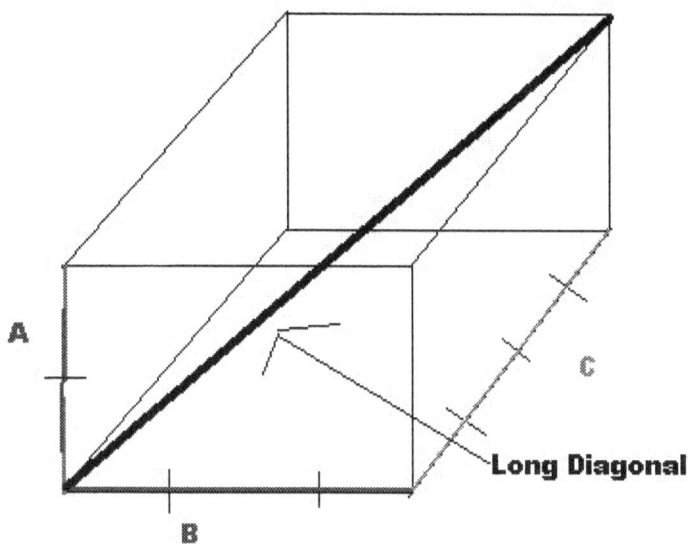

The sides are labeled A, B, and C. The line between the two corners represents the long diagonal of the box.

The formula for the long diagonal's length is the square root of A squared plus B squared plus C squared. That would be the square root of 4 plus 9 plus 16 which equals the square root of 29.

Pictures of One Setup of Instrument

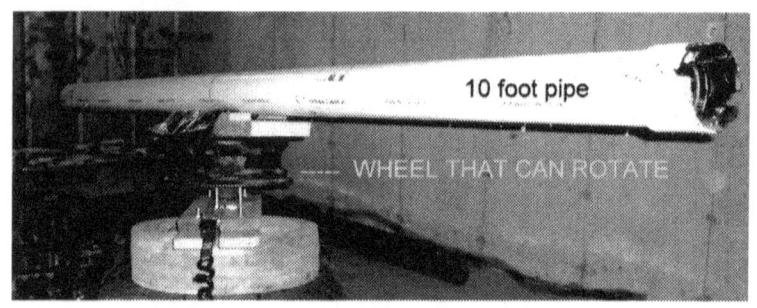

10 foot pipe

WHEEL THAT CAN ROTATE

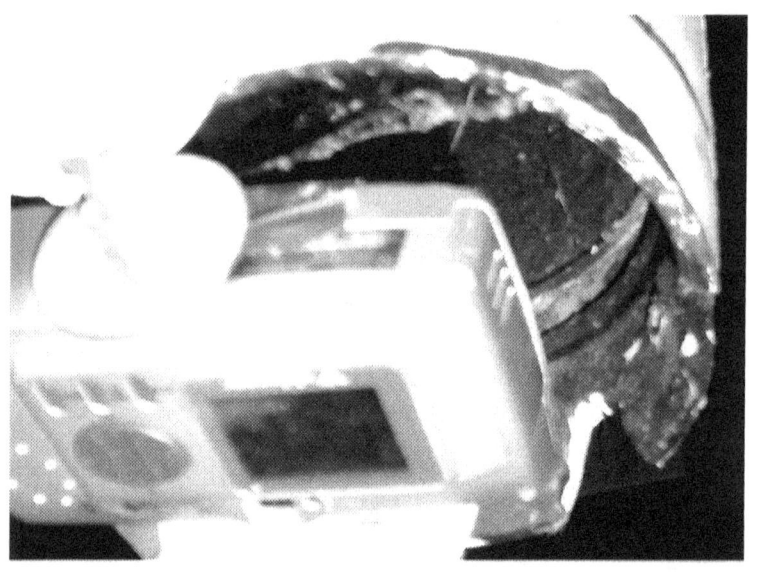

CAMERA

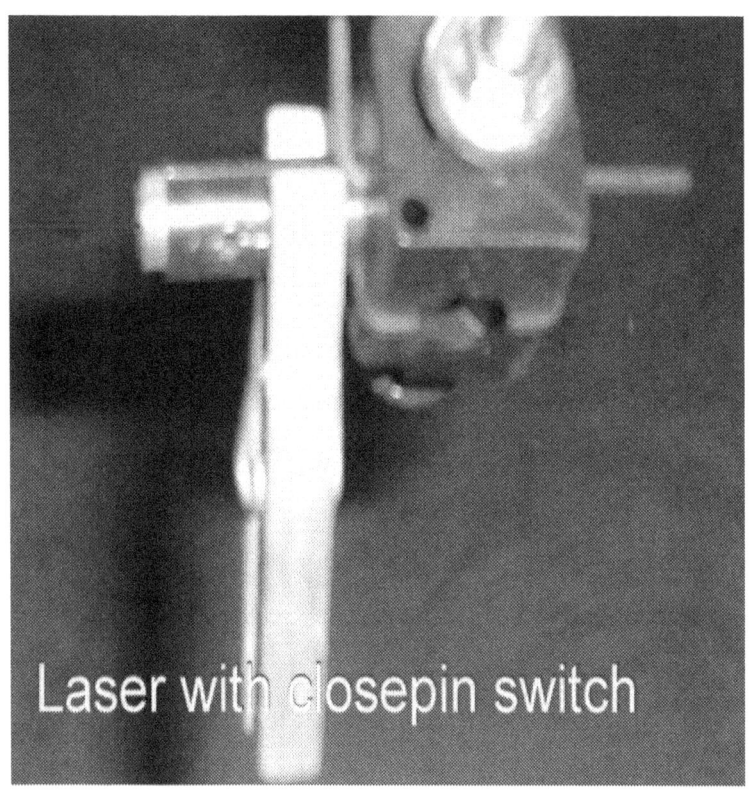

Laser with closepin switch

Converting Micron per Nanoseconds to Kilometers per Second.

My setup measured how far a CCD would move in the time it takes light to travel 10 feet. There are 1,000,000,000 nanoseconds in one second. Scientific Notation is 10 to the power of 9. So to convert 10 nanoseconds to 1 second I have to multiply by 10 to the 8, which is 100,000,000.

If the CCD moved about 300 micron in the 10 nanoseconds, this would mean it would move in 1 second 300 times 100,000,000 microns.

That would make 30,000,000,000 micron.

Now to convert this number to kilometers we do the following. Divide by 1,000 and that gives 30,000,000 millimeters. Divide by 1,000 to get to meters and that is 30,000. Then to get to kilometers divide by 1,000 and that provides 30 kilometers. This then is 30 kilometers per second. Of course if one knows Scientific Notation that makes life easier.

Also an easier way is just to take the movement noted on the CCD which is 300 and divide that by 10 to get 30 Klicks (Kilometers) per Second.

To go to miles per second multiply Klicks per second by .62. (There are .62 miles in a kilometer.) This would be 18.6 miles per second. Then to convert to miles per hour, multiply by 3600, the number of seconds in an hour. This is then 66,960 miles per hour.

Reporter:

I would like you to provide another example of how you view the earth's vector of motion and how you detect it. Sorry to be so insistent but I want to be sure I have it in my mind.

Author:

All right, let me have you imagine that the vector is an arrow actually visible and in the center of a room. It has length and that represents the speed of the motion. A longer arrow means more speed and a shorter arrow means less speed. The direction the arrow is pointing is the direction the motion is pointing. The combination of speed and direction define the vector of the earth's motion. Now take our camera and photograph the arrow. Move around the arrow in a circle stopping every so often and take another picture. When you are facing the back of the arrow and the front is directly away from you the arrow will be just a dot on the picture. When you are facing the arrow from the side you will see the total length of the arrow. This view provides the tip and feathers on the back of the arrow. Thus as you walk around the arrow taking pictures the size of the arrow on the developed pictures will vary from a dot to the full length back to a dot then the full length again. The view depends on where you are as you take the picture.

This is similar to what the Velador does in capturing the earth's vector of motion. Unfortunately the actual earth's vector is not still. It is moving and changing where it points all the time as viewed from an Earth site. And I suspect the arrow is curved as well. So it is actually very difficult to capture in its entirety. Despite this there are many opportunities to study what is happening and interpret the results. It is my belief that there is a very lot more science wrapped up in this topic and it awaits exploration and explanation.

Reporter:

Thanks that seems to jell with what you have said previously. Any thing else?

Author:

Yes there is. I was not going to mention this but believe I really should. In the construction of the Velador one should gently remove the lens of the camera without disturbing the other electronic functions of the camera. The laser light should impinge upon the CCD directly with out the influence of the lens in place. One can use a bit of tape to cover the opening when the camera is not in use so dust and dirt do not settle on the CCD. On my setup I have held my breath each and every time I have used it but the camera has held up remarkably well. I have to say very remarkably well. And there is a bit more to the thought experiment but that is up to the problem solver to figure that part out.

Note Page

Note Page

Note Page

About The Author

Retired medical doctor pursues interest in physics topic after reading about a Victorian era experiment. Both a novice in physics and book publishing, he presses forward.

www.ingramcontent.com/pod-product-compliance
Lightning Source LLC
Chambersburg PA
CBHW022022170526
45157CB00003B/1320